Reducing thermal bridging at junctions when designing and installing solid wall insulation

Caroline Weeks, Tim Ward and Colin King

The research and writing for this publication has been funded by BRE Trust, the largest UK charity dedicated specifically to research and education in the built environment. BRE Trust uses the profits made by its trading companies to fund new research and education programmes that advance knowledge, innovation and communication for public benefit.

BRE Trust is a company limited by guarantee, registered in England and Wales (no. 3282856) and registered as a charity in England (no. 1092193) and in Scotland (no. SC039320).

Registered office: Bucknalls Lane, Garston, Watford, Herts
WD25 9XX
Tel: +44 (0) 1923 664743
Email: secretary@bretrust.co.uk
www.bretrust.org.uk

IHS (NYSE: IHS) is the leading source of information, insight and analytics in critical areas that shape today's business landscape. Businesses and governments in more than 165 countries around the globe rely on the comprehensive content, expert independent analysis and flexible delivery methods of IHS to make high-impact decisions and develop strategies with speed and confidence. IHS is the exclusive publisher of BRE Trust publications.

IHS Global Ltd is a private limited company registered in England and Wales (no. 00788737).

Registered office: Willoughby Road, Bracknell, Berkshire
RG12 8FB.
www.ihs.com

BRE Trust publications are available from:
www.brebookshop.com
or
IHS BRE Press
Willoughby Road
Bracknell
Berkshire RG12 8FB
Tel: +44 (0) 1344 328038
Fax: +44 (0) 1344 328005
Email: brepress@ihs.com

Printed using FSC or PEFC material from sustainable forests.

FB 61

First published 2013

ISBN 978-1-84806-350-1

Cover photographs:
Typical external wall insulation being applied (left)
Thermal bridging modelling image of an external wall/ground floor junction (top right)
Thermal image of a dwelling (bottom right; courtesy of Joanne Hopper)

Contents

Executive summary — iv

1 Introduction — 1

2 Background — 2

2.1 Risks associated with thermal bridging — 3
2.2 Solid wall constructions modelled — 3
2.3 Key junction details modelled — 4

3 Detailing of external wall insulation — 5

3.1 Window jamb — 5
3.2 Window head — 6
3.3 Window sill — 7
3.4 Eaves — 8
3.5 External wall/ground floor junction — 9
3.6 Party wall/external wall junction — 10
3.7 Quality control on site — 11
3.8 Conclusions for external wall insulation — 11

4 Detailing of internal wall insulation — 13

4.1 Interstitial condensation risk — 13
4.2 Window jamb — 13
4.3 Window head — 14
4.4 Window sill — 16
4.5 Intermediate floor/external wall junction (within the same dwelling) — 17
4.6 Intermediate floor/external wall junction (in apartments) — 18
4.7 Party wall/external wall junction — 19
4.8 Quality control on site — 20
4.9 Conclusions for internal wall insulation — 20

5 Comparing the overall thermal performance of external and internal insulation — 21

5.1 Conclusions for external versus internal wall insulation of solid walls — 21

6 References — 24

Executive summary

With the advent of the Green Deal, Energy Company Obligation (ECO) and other financial incentives to improve the UK's existing building stock, there has never been a more relevant time to push the construction industry towards better performance. Current practice during refurbishment does little or nothing to minimise the effects of thermal bridging or inconsistency in thermal envelope performance. This guide sets out clear principles and methods that should be considered and adopted during the design and installation of solid wall insulation in order to reduce thermal bridging effects, maximise carbon dioxide (CO_2) emission reductions and minimise the risk of condensation.

The effect of installing external and internal wall insulation in typical solid wall homes has been modelled for junctions with windows, eaves, floors and party walls. Potential problems are considered, taking examples from recent refurbishment projects in which BRE has been involved. This BRE Trust Report will be a useful resource for public and private clients looking to improve the performance of their properties and for architects/designers, specifiers and installers.

1 Introduction

Improving the energy efficiency of the existing building stock is one of the biggest challenges facing the UK. In particular, traditional solid wall houses are more difficult and more costly to improve than more modern, cavity wall constructions. Initiatives such as the Green Deal should serve to finally encourage such refurbishment. However, current industry practice does little or nothing to minimise the effects of thermal bridging or inconsistency in thermal envelope performance when installing insulation in solid wall dwellings. Making the effort to minimise thermal bridges is considered likely to add more time and expense to what is already regarded as a costly improvement measure.

This BRE Trust Report seeks to highlight the importance of appropriate detailing for both externally and internally applied solid wall insulation and demonstrates its effect on heat flow and potential condensation risk at key junctions.

2 Background

Improving the thermal efficiency of solid wall properties by installing externally or internally applied insulation is becoming an increasingly popular measure to cut energy bills for occupants and to reduce CO_2 emissions. Either method of insulating can bring about significant reductions in heat loss to a dwelling, and there are various reasons why one method may be favoured over the other. Some of the key considerations are highlighted in Table 1.

Both insulation methods can introduce potential problems due to thermal bridging at key junctions if the detailing is not carefully considered. This report looks at all relevant junctions for each method of retrospectively applied insulation where typical practice could exacerbate the effects of thermal bridging; namely (i) increased heat loss from a dwelling compared with what was assumed during energy modelling; and (ii) the risk of condensation or mould growth from lowered internal surface temperatures at or near to the thermal bridge junction.

Table 1: Advantages and disadvantages of external and internal wall insulation

Considerations	External wall insulation	Internal wall insulation
External appearance	Can improve the rainscreen protection and appearance of property and can therefore enhance the property value.	No change in external appearance, which may be a benefit when wishing to retain attractive external features.
Influence on floor areas	Some loss of space externally due to insulation (may affect pathways). Surface may be less robust than solid wall, which may require reinforcement in vulnerable areas.	Some loss of internal floor area/room size. Surface may be less robust to knocks or damage than solid surface.
Removal/refitting of items	Items such as guttering/drainage and aerial/satellite dishes may need to be moved or refitted to accommodate insulation.	Services, skirting, radiators, switches and plug points will need to be removed and refitted onto newly insulated surface. Subsequent redecorating will also be required.
Continuity of insulation	Generally good continuity across plane wall elements, though there may be some disruptions to the insulation layer from adjoining garden walls, garages and lean-to structures.	There will inevitably be breaks in the insulation continuity at internal partition walls and intermediate floors.
Thermal mass and responsiveness	Reduction of the swings in internal temperature, eg from solar gain, due to the increase in thermal storage of the building fabric.	Increase in the swings in internal temperature, but an increase in responsiveness to the heating system.
Disruption to household	Scaffolding will be required to access upper storeys, but otherwise minimal disruption to occupants.	Would need to move furniture and fittings and may need to cut off services temporarily to some locations. Easiest if dwelling is empty and unoccupied at time of works.
Planning requirements	Planning permission may be required if the new insulation will front directly onto a public right of way or the building is in a conservation area or similar.	Since the external appearance will not be changed, no permissions should be required.
Protecting the wall	Can extend the lifespan of the building by protecting the brickwork/stonework, eg from driving rain.	Offers no weather protection to the wall.
Potential for partial works	Needs to be done over an entire facade at least, if not the entire dwelling.	Can be done on a room-by-room basis if necessary.

A 'thermal bridge' is the informal term given to any part of the building envelope where there is an increased heat flow compared with the adjacent parts. In the context of the thermal bridging associated with junctions formed between building elements or around openings, this additional heat loss is the linear thermal transmittance – ψ (pronounced 'saI') – associated with the length of the particular junction.

There are two thermal bridging effects that can contribute to the linear thermal transmittance of a junction detail and both effects are often present together. One is a geometric effect that results from internal and external areas at the junction not being equal. The other effect is a constructional effect, eg where the insulation thickness or thermal conductivity changes or where the insulation is penetrated by materials of higher thermal conductivity.

Values of linear thermal transmittance (ψ-values) apply over the length of the particular junction, such as the perimeter of roof eaves. Some thermal bridges result in 'point thermal transmittance', ie the additional heat loss applies to a single point, as represented by χ-values (pronounced 'kaI'). Overall, the total heat loss from any building will be determined from:

- U-values multiplied by the area over which they apply (W/m²K), plus

- Ψ-values multiplied by the length over which they apply (W/mK), plus

- χ-values multiplied by the number of occurrences of the particular point bridge(s) (W/K).

2.1 Risks associated with thermal bridging

Thermal bridging at junctions or around openings will give rise to additional (ie unaccounted for) heat loss. If this additional heat loss is ignored (ie assumed to be zero) when calculating the total heat loss through the fabric of the building, it is likely that the overall heat loss will be underestimated by a greater or lesser degree, depending on the level of thermal bridging that exists at junctions. Thus, if buildings are improved through insulation of the plane building elements (such as the wall) but thermal bridging at the junctions is either ignored or not properly determined, the heat loss may be significantly higher than intended.

As an example of this phenomenon, Figure 1 is a thermal image that gives a visual representation of the thermal bridging effect at window and door reveals when externally applied wall insulation is not continued into the reveals. This particular type of junction is discussed in detail in Section 3.1 of this report.

Individual junctions that have not been properly considered can cause temperatures at the inside surface of the insulated property (or in some cases also in a neighbouring uninsulated property at party walls or floors) to become so low that there is a risk of surface condensation or mould growth. In the context of thermal bridging at junctions, the reduced temperature that occurs at the inside surface at or near a thermal bridge is expressed as a 'minimum temperature factor', which is a property of the thermal bridge and its construction. In order to avoid the risk of surface condensation or mould growth at junctions, the minimum temperature factor should be greater than or equal to the critical value (f_{CRsi}) of 0.75 , as quoted

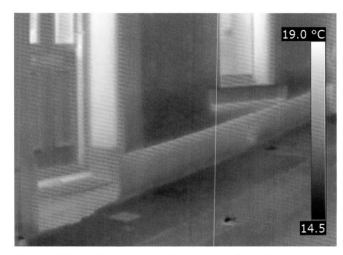

Figure 1: Thermal image of a dwelling with typically applied external wall insulation, where the window and door reveals have remained uninsulated and hence show significant relative heat loss (Courtesy of Joanne Hopper)

in BRE Information Paper IP 1/06[1]. It is therefore important when assessing the effects of thermal bridging at a junction to also understand the potential impact on the inside surface temperatures at or near thermal bridge junctions.

2.2 Solid wall constructions modelled

The modelled solid wall constructions in this study are of solid brick walls. However, the overall conclusions apply to the wider range of solid walls present in the UK, since the methods of insulating either externally or internally are relatively standardised, irrespective of the precise nature of the solid walling. Since the most likely reason for considering internal rather than external wall insulation is where decorative stone or brickwork frontages are present on a building, it is these types of walling where internal insulation will be most relevant.

A simplified construction has therefore been assumed for the modelling, which nevertheless demonstrates the general trends that are likely to be witnessed for all solid walls. Since the constructions modelled are simplified, the values quoted in this report should be treated as indicative and so used for illustrative and comparative purposes only. If values are required for a specific construction then bespoke modelling should be undertaken.

The simplified construction models assume a generic solid brick wall with externally applied sand render and a plastered internal finish. Where sills are included these are taken to be of limestone, and lintels are assumed to be concrete. External or internal wall insulation with a thermal conductivity of 0.03 W/mK, such as expanded polystyrene (EPS), is used throughout the models. Where junctions include loft insulation this is assumed to have a thermal conductivity of 0.04 W/mK, such as mineral wool. Table 2 lists all assumed thermal conductivities for each of the key materials. The theoretical 'typical' improvement scenarios detailed here have been based on experience of such installations and on manufacturers' standard details. The external insulation details in particular are in line with guidance in *The complete guide to external wall insulation*[2], which is widely used as an industry reference source.

This thermal bridging modelling has been carried out using Physibel's Trisco (v12) software, following the guidance and conventions in BRE Report BR 497[3], as required for compliance with UK building regulations. The minimum temperature factor (f_{Rsi}) for each junction is also quoted. It should be noted that in each case the internal temperature of the modelled scenario is taken to be 20°C and the external temperature 0°C (ie 'standard conditions').

Table 2: Material property assumptions used in the models

Material	Assumed conductivity (W/mK)
Brickwork	0.770
Cement sand render	1.000
Plaster	0.400
Concrete lintel	1.500
Limestone sill	1.700
Concrete floor	2.000
Hardcore/ground	1.500
PVC-U drip or capping plate	0.170
Hardwood sill board	0.180
Softwood header plate, fascia or skirting board	0.130
Mineral wool insulation (loft)	0.040
EPS external insulation	0.030
Phenolic insulation	0.020
Narrow airspace for damp-proof course in ground floor detail (calculated to BS EN ISO 6946:2007[4])	0.398

2.3   Key junction details modelled

The modelling results for the key junctions that apply to the solid walls being insulated externally are considered in Section 3 of this report and those that apply to the solid walls being insulated internally are considered in Section 4.

These key junctions are those whose thermal performance is most influenced by the application of external or internal wall insulation to uninsulated solid walls. They have been selected and modelled to demonstrate their impact on both the heat loss (ψ-values) and the minimum temperature factor (f).

For each key junction, the following sets of figures show:

    a.   the materials in the uninsulated junction detail

    b.   colour gradations indicating how the temperature changes through the uninsulated detail (red denotes the higher temperatures, through to blue, which represents the colder temperatures)

    c.   the temperature changes through the 'typically insulated' junction

    d.   the temperature changes through the 'improved' junction.

Each of the colour-graded images also show:

- a set of heat-flow lines (in the image) indicating the density of the heat flow through the detail (where each line represents the same heat loss of 2 W)

- the ψ-value and the temperature factor for the junction.

The most vulnerable part of the junction, ie the part of the internal surface with the lowest temperature, can be readily identified in the colour-graded images, along with the densest, most concentrated areas of heat flow (where the heat-flow lines are closest together). For each of the scenarios, the quantified thermal performance, ie the ψ-value and the minimum temperature factor, can be immediately compared and any differences readily highlighted.

3 Detailing of external wall insulation

3.1 Window jamb

For the window jamb junction in the 'uninsulated' base case, it is assumed that the window frame is situated in the centre of the wall cross-section, with external render and internal plaster returned to the window frame (Figure 2a). In the case where the wall is externally insulated, it is assumed that the insulation will run flush to the edge of the jamb and the render finish will again be returned to the window (Figure 2c).

To improve this junction, it is recommended that a fillet of insulation is applied within the window reveal prior to rendering (Figure 2d). In this 'improved' case, a 20 mm thickness of insulation with a thermal conductivity of 0.02 W/mK (such as phenolic foam) is assumed, since typically there will be a 30–40 mm space between the existing wall and any opening casement windows that may be present. In reality, the thickness of the applied insulation has to be tailored to the available space within the reveal, and a lack of such space is often quoted as the reason why improved detailing cannot be practically achieved.

Such restrictions are lessened if new windows are installed at the same time as the external insulation is applied. In such circumstances it is then possible to design appropriate window frame arrangements to allow the desired insulation detail to be included. There may also be scope to insulate across the entire cross-section of the wall at the reveal, but this is not explored in this study. This issue regarding replacement windows applies also to the window head and window sill junctions that follow.

The application of the external insulation to the wall significantly reduces the U-value of the wall, hence reducing the overall heat loss. However, the concentration of different-coloured bands around the location of the window frame in Figure 2c indicates the rapid temperature drop across this particular location. The ψ-value of Figure 2c is significantly higher than the base-case ψ-value in Figure 2b, the reason being that although the heat loss through the wall itself has been reduced, there is more lateral heat loss flowing towards the junction.

It is interesting to note that, despite this concentration of heat flow at the junction, the temperature factor at the coldest internal point of the junction (where the plaster meets the window frame) has in fact improved from the base case and does not drop below the critical value of 0.75.

Figure 2d includes additional insulation returned into the reveal. Here the ψ-value has returned (ie reduced) to a more acceptable value and the temperature factor has risen again, hence further reducing the risk of condensation or mould growth at the junction.

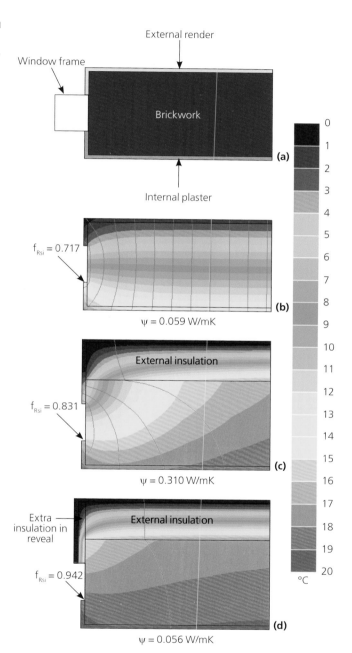

Figure 2: Junction detail for window jamb
a: 'Uninsulated' detail
b: 'Uninsulated' modelling results
c: 'Typical' external insulation
d: 'Improved' external insulation

3.2   Window head

For this junction, it is assumed that a concrete lintel is present across the head of the window and that the window frame is again situated in the centre of the wall cross-section, with external render and internal plaster returned to the window frame (Figures 3a and 3b). When the wall is externally insulated, it is assumed that the insulation will run flush to the edge of the lintel and the render finish will again be returned to the window frame (Figure 3c).

To improve this junction, it is recommended that a fillet of insulation is applied within the window reveal at the head prior to rendering (Figure 3d). In this 'improved' case, 20 mm of insulation with a thermal conductivity of 0.02 W/mK, such as phenolic foam, is assumed.

As with the jamb junction, the application of external insulation to the wall without insulating the reveal (Figure 3c) reduces the heat loss overall, but causes an increase in the ψ-value (compared with the 'uninsulated' base case of Figure 3b) to 0.445 W/mK. The temperature factor at the coldest point on the internal surface of the junction (where the plaster meets the window frame) has increased from that in the 'uninsulated' base case to 0.800 and is now above the critical value of 0.75, indicating no risk of surface condensation or mould growth.

Figure 3d shows the 'improved' case where additional insulation is returned into the reveal at the window head. Here the ψ-value is significantly reduced to 0.064 W/mK, which is lower than both the 'uninsulated' base case and the 'typical' insulation case. The temperature factor has further improved to 0.944, indicating no risk of condensation or mould growth at the inside surface of the junction. Note that if the concrete lintel is replaced by a traditional timber lintel (that is less thermally conductive) the risk of thermal bridging at this junction is reduced significantly.

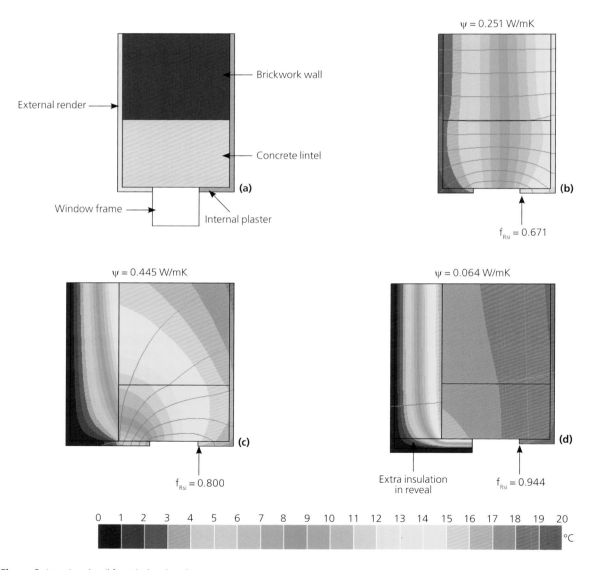

Figure 3: Junction detail for window head
a: 'Uninsulated' detail; **b:** 'Uninsulated' modelling results; **c:** 'Typical' external insulation; **d:** 'Improved' external insulation

3.3   Window sill

Here it is assumed that a traditional limestone sill spans the wall at this junction and that the window frame is again situated in the centre of the wall cross-section. External render is applied up to the underside of the stone sill and a hardwood inner sill board finishes the internal reveal (Figures 4a and 4b). When the wall is externally insulated, it is assumed that the insulation would be brought in front of the stone sill but finish level with the sill surface. A secondary extended drip-sill (probably PVC-U) would then be applied to the window opening so as to direct water over the newly applied insulation and clear of the wall (Figure 4c).

To improve this junction a fillet of insulation is applied to the external sill, which can subsequently be finished with an extended PVC-U drip-sill (Figure 4d). In this case, 20 mm of insulation with a conductivity of 0.02 W/mK, such as phenolic foam, is assumed underneath the external PVC-U sill.

As with the other window junctions, the application of external insulation to the wall (Figure 4c) without insulating the reveal does significantly reduce the overall heat loss, but causes an increase in the ψ-value to 0.483 W/mK, compared with the 'uninsulated' base case of 0.186 W/mK (Figure 4b). The temperature factor at the coldest point on the internal surface of the junction (at the edge of the timber internal sill board adjacent to the window frame) has in fact improved to 0.789 from that of the base case of 0.662, but is uncomfortably close to the critical value of 0.75, indicating a slight risk of surface condensation or mould growth.

Figure 4d shows the 'improved' case, which has additional insulation returned into the reveal above the stone sill. As a result the ψ-value is reduced to 0.110 W/mK, which is below both that of the 'uninsulated' base case and that of the 'typical' insulation case. The temperature factor is 0.925, which has improved to significantly above the critical value of 0.75, indicating no risk of condensation or mould growth.

Note that it would be possible to further improve this junction by cutting back the extruding stone sill flush with the brick wall, continuing the external insulation up in front of this and then applying the additional fillet of insulation on the top surface of the sill. This would reduce the ψ-value by a significant amount, to 0.059 W/mK.

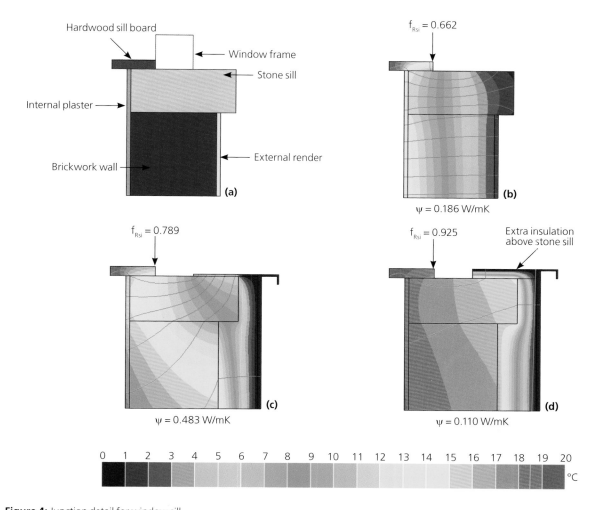

Figure 4: Junction detail for window sill
a: 'Uninsulated' detail; **b:** 'Uninsulated' modelling results; **c:** 'Typical' external insulation; **d:** 'Improved' external insulation

3.4   Eaves

At the eaves, it is assumed that a timber head plate would run across the top of the brick wall, to which the rafters and trusses would be fixed (not shown in the diagram). The inner wall surface is again finished with plaster and the external surface with sand render. A timber fascia board is applied at the top of the wall and a soffit box with rainwater goods attached to it. The ceiling is assumed to be of lath and plaster-type construction with a 200 mm thickness of mineral insulation above, since most households installing wall insulation will have already installed loft insulation. The 'uninsulated' base case is shown in Figures 5a and 5b.

When the wall is externally insulated it is assumed that the insulation would be brought up to the edge of the eaves box and capped with a PVC-U profiled flashing plate to prevent moisture being able to penetrate behind the insulation layer (Figure 5c). Although a costly measure, it is possible to improve the eaves detail and allow continuity of the insulation at this junction by extending the eaves so that the loft insulation can be brought out across the top of the externally applied wall insulation (Figure 5d).

The ψ-value for the 'uninsulated' base case is slightly negative at -0.010 (Figure 5b). This is due to the timber at the wall head, which has a lower thermal conductivity compared with that of brick, replacing the brickwork, ie the U-value is applied to an internal area where part of the external wall at the top has a greater thermal resistance compared with that of the rest of the wall. However, the minimum temperature factor of 0.742 is slightly below the critical value of 0.75, indicating a slight risk of surface condensation or mould growth at the internal surface of the junction.

In the 'typical' insulation case (Figure 5c), the external wall insulation stops short of the height of the wall (as seen from the inside) by virtue of the eaves box and thus there is a significant area at the top of the wall that is not insulated. The additional heat loss that results is then attributed to the ψ-value of the junction, which is significant at 0.331 W/mK. However, interestingly the inside temperatures have improved such that the minimum temperature factor is now significantly above the critical value of 0.75 and so there is no risk of surface condensation or mould growth for this case.

For the 'improved' insulation case (Figure 5d) where the loft insulation is brought out to overlap the extended external insulation, there is no path for heat to flow into the wall construction and then bypass the external wall insulation. Consequently, the ψ-value is significantly reduced to 0.042 W/mK, compared with the 'typical' insulation case. The minimum temperature factor further increases to 0.945 giving no risk of condensation or mould growth for this junction.

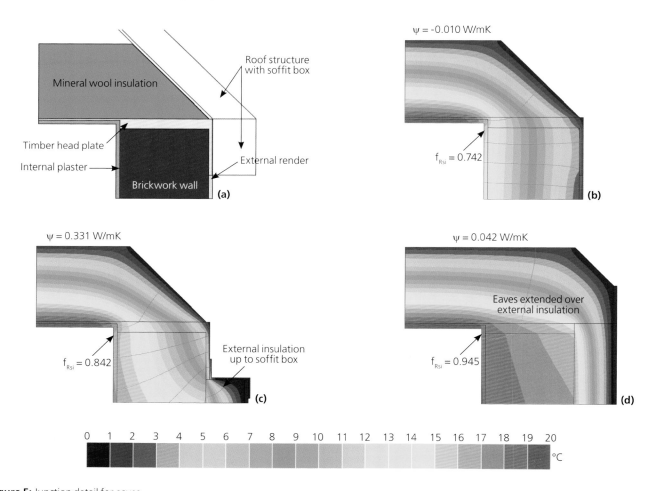

Figure 5: Junction detail for eaves
a: 'Uninsulated' detail; **b:** 'Uninsulated' modelling results; **c:** 'Typical' external insulation; **d:** 'Improved' external insulation

3.5 External wall/ ground floor junction

It is assumed that the solid brick wall sits directly on a strip foundation, with an internal plaster finish with softwood skirting board and an external sand render finish, stopping approximately 150 mm above ground level. The floor comprises in-situ concrete laid onto compacted hardcore (Figures 6a and 6b). When the wall is externally insulated (Figure 6c), it is usual for the insulation to stop clear of the ground to preserve the damp-proof course and to be supported by a galvanised steel baseplate.

In order to overcome the obvious and significant thermal bridge at the section of uninsulated brick wall, it is clear that the external insulation needs to be extended down to (or below) ground level (Figure 6d), while retaining a small channel/break in the insulation to prevent moisture from being drawn up behind the insulation by capillary action. This channel would need to be appropriately sealed at the channel opening to prevent any moisture ingress from the external surface.

Applying external insulation to the wall (Figure 6c) but retaining approximately 150 mm height of uninsulated brickwork gives rise to a significant additional heat loss, which is then attributed to the ψ-value. As a result the ψ-value is a significant 0.604 W/mK compared with the 'uninsulated' base case (Figure 6b) of 0.149 W/mK. The temperature factor at the coldest point on the inside surface of the junction (ie on the internal wall surface where it meets with the skirting board) is 0.700 and is almost unchanged compared with the base case of 0.691. Note that for both the 'uninsulated' base case and the 'typical' insulation case, there is a risk of condensation and mould growth on the internal surface of this ground floor junction.

In the case of the 'improved' junction (Figure 6d), where the external insulation is taken to ground level the ψ-value has reduced significantly to 0.154 W/mK. The minimum temperature factor has improved significantly to 0.805, which is now above the critical value of 0.75, indicating no risk of condensation or mould growth at the inside surface of the junction. If practical, extending the external insulation to at least 150 mm below the ground level would reduce the ψ-value for this junction further to 0.016 W/mK (this case is not shown).

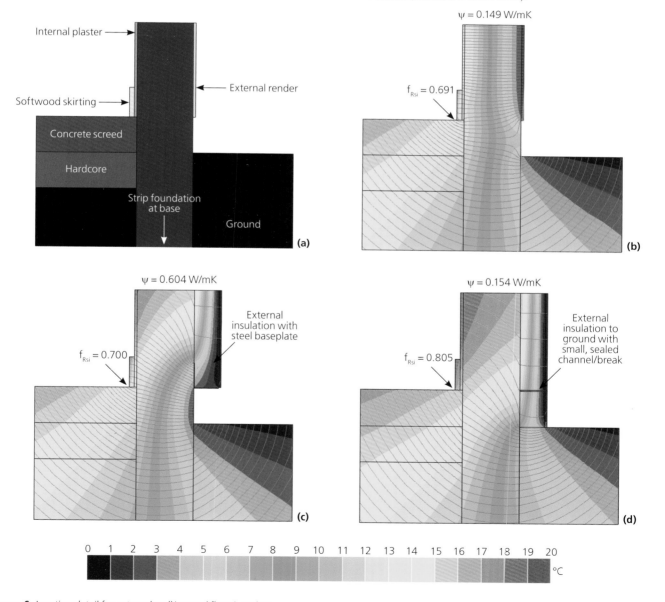

Figure 6: Junction detail for external wall/ground floor junction
a: 'Uninsulated' detail; **b:** 'Uninsulated' modelling results; **c:** 'Typical' external insulation; **d:** 'Improved' external insulation

3.6   Party wall/external wall junction

Since the majority of dwellings in the UK are attached to a neighbouring dwelling (ie not detached), it is relatively likely that situations will arise when one property in a neighbouring pair receives externally applied wall insulation, while the other does not. For this situation it is assumed that the external wall and the adjoining party wall are of brickwork construction, with an internal plaster finish and an external sand render finish (Figures 7a and 7b). It is also assumed that when one of the adjoining dwellings receives externally applied insulation, it will be taken to the mid-point of the party wall and finished with render (Figure 7c). The situation where both dwellings are insulated externally is given in Figure 7d.

For the 'uninsulated' base case (Figure 7b), the ψ-value for the junction is 0.465 W/mK, and since the heat flow is almost symmetrical across the junction half is allocated to each dwelling. However, in the 'typical' insulation case, where one property is insulated and the other not (Figure 7c), heat loss from the insulated dwelling is much reduced compared with the base case, while that for the uninsulated dwelling is almost unchanged. As a consequence, however, the heat flow across the junction from one property compared with the other is now very asymmetric. Because of this, separate ψ-values for

this junction must be determined for each property. In this case, instead of the total ψ-value (0.474 W/mK) being halved, the insulated property has a ψ-value of 0.329 W/mK and the uninsulated property a value of 0.146 W/mK.

Regarding the minimum temperature factors, the coldest surface temperature on the insulated side is much reduced with a minimum temperature factor of 0.977, indicating no risk of condensation or mould growth at the inside surface of the junction in the insulated property. Interestingly for the uninsulated property, rather than the temperature factor decreasing compared with the 'uninsulated' base case, as may instinctively be expected, it also improves, but only marginally to 0.742. This temperature factor is only just below the critical temperature factor, indicating a slight risk of interstitial condensation or mould growth, but importantly this risk is less than in the base case of 0.730, ie prior to the neighbouring property being insulated. This means that for the party wall junction with the external wall, insulating one property and not the other has no negative impact on the uninsulated property and, indeed, for this junction, insulating one property and not the other results in some reduction in heat loss from the uninsulated property. Note that this junction is significantly improved if both properties are insulated. If this is done, the ψ-value for the whole junction is reduced to 0.064 W/mK, with 0.032 W/mK allocated to each dwelling (Figure 7d).

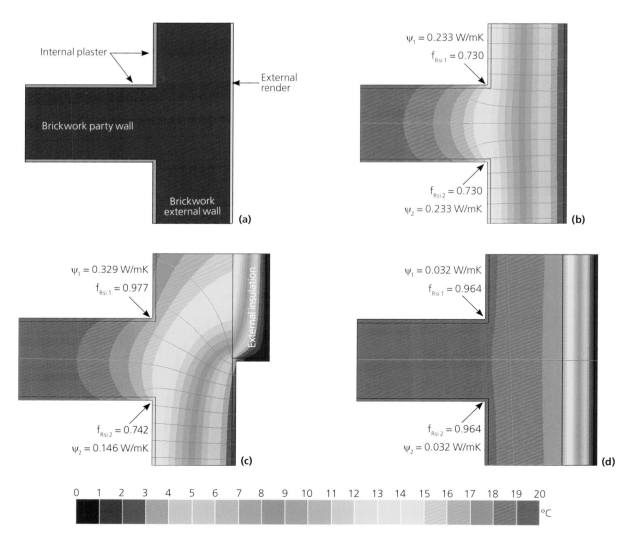

Figure 7: Junction detail for party wall/external wall junction
a: 'Uninsulated' detail; **b:** 'Uninsulated' modelling results; **c:** 'Typical' external insulation (to only one dwelling); **d:** External insulation (to both dwellings)

3.7 Quality control on site

When fitting external wall insulation, attention to detail during installation is vital. This is not just to ensure the finished work is aesthetically pleasing, but also to ensure that additional unintended thermal bridges are not built into the structure by poor workmanship or inadequate detailing of junctions/connections other than those covered in this study. This section gives some examples of these, as witnessed in real site installations.

3.7.1 Detailing around obstructions

Consideration needs to be given to how insulation will be detailed at potential obstacles on the external facade, such as incoming service mains, telephone junction boxes, utility boxes, satellite dishes and adjoining items such as fences or walls that are in such a position as to hinder the installation process. Continuity of the insulation layer should be the ultimate aim so that, wherever possible, obstacles should be removed and replaced to allow insulation to continue uninterrupted. If this cannot be done, the areas where insulation is not present will effectively allow heat to bypass the insulation layer, resulting in increased heat loss at these locations.

Depending on the relative heat loss area, this may or may not make a significant contribution to the overall heat loss of the building. However, more significantly, it may result in the temperature factor being below the critical limiting value at these locations, leading to an increased risk of condensation and mould growth at the inner wall surfaces of these uninsulated areas. Clearly this would be undesirable, so potential obstructions and irregular detailing should be identified during the design stage of each particular building insulation programme and measures to eliminate such potential problem areas should be proposed.

Figure 8 shows where external wall insulation has been cut around external pipework entering the dwelling. Clearly this area will result in concentrated heat loss from the property, particularly since metal 'capping plates' have been used at the exposed edges, which will be very conductive relative to the other surrounding materials. A preferable solution would have been to temporarily disconnect the pipework, provide an extending section to reach beyond the line of the newly applied insulation, then to reconnect the pipe after installation. This procedure potentially involves additional trades, depending on the nature of the pipe being disturbed, and may result in some degree of inconvenience to the household – factors that will inevitably act as deterrents to more favourable detailing. However, the fact remains that under certain conditions this detail could cause potential problems of condensation and mould growth on the inner wall surface behind the uninsulated section of wall.

3.7.2 Workmanship

For the window junctions covered previously, it is suggested that a different, higher performance insulation material is used to fill the small available gaps within reveals instead of that used for the insulation of the plane elements. This maximises the benefit that can be achieved by applying this detail. The benefit would be reduced if another insulation product were selected on site – particularly a more poorly performing material. It should be emphasised to installers that it is very important to use the correct insulation material for the place where it is specified.

Additional unintended thermal bridges can also be introduced by poor workmanship that causes inconsistency in the applied

Figure 8: Pipework not extended for this installation, preventing the continuity of externally applied wall insulation, which will lead to unforeseen thermal bridging effects

Gap introduced between reveal insulation and main wall insulation due to poor workmanship

Figure 9: Window jamb detail with external wall insulation, showing gaps between the main insulation to the wall and the smaller insert of insulation within the window reveal

insulation. An example is poor butting-up/alignment of insulation, which creates small gaps in the insulation layer. Care should be taken when detailing around openings, particularly where the type of insulation changes at reveals, to ensure no unintended gaps are introduced that could negate the benefit of the improved detail being delivered. Figure 9 shows an installation where gaps have been introduced between EPS external wall insulation and the phenolic insulation applied within the window reveal. This detail is intended to significantly reduce thermal bridging at this junction, but such air gaps will allow heat to bypass the insulation and escape from the building more readily. An individual instance of this is unlikely to result in significant heat loss from the building, but poor installation in a number of places across a dwelling could do so collectively.

3.8 Conclusions for external wall insulation

These results from modelling the key junction details for external wall insulation have demonstrated that thermal bridging becomes increasingly significant when insulation is applied externally to the solid walls of dwellings using typical industry practices. However, this increase in thermal bridging can generally be mitigated by giving special attention to the precise detailing employed. When designing the detailing around junctions before installing external insulation, consider the following:

- Attention should be given to applying insulation within the reveals at window junctions and especially to isolating thermally conducting features such as concrete or stone lintels or sills.

- Where external walls meet the ground floor, it is possible to reduce the extent of thermal bridging by continuing the external insulation to ground level (or below), with proper consideration given to preserving an appropriate damp-proof mechanism.

- Eaves junctions may be improved by extending the eaves so as to allow the loft insulation at ceiling level to be brought out to overlap that of the external wall insulation, which will also require extending up to meet the ceiling insulation. However, this is a more complex and costly solution compared with some of the others discussed in this study and will need to be considered on a case-by-case basis, depending on the practical circumstances of any particular installation, eg the effect on neighbouring properties.

- At the junction of party walls with external walls, the modelling suggests that insulating only one dwelling in an adjoined pair will not detrimentally affect the uninsulated dwelling. However, the best way to improve the thermal bridging at such a junction is to insulate both dwellings.

Although thermal bridging at junctions may be exacerbated as a result of poor detailing with externally applied insulation, it is reassuring to note that applying external insulation did not make the minimum temperature factors any worse for the junctions investigated in the modelling. Hence there should in general be no significant increase in the risk of condensation or mould growth. Of course some households, as a result of poor internal conditions of low temperature and/or high humidity, may still experience condensation and mould growth, but, although these might first appear and be worse at the inside surfaces near junctions, this in general will not be as a result of the installation of external insulation.

It is important that external wall insulation is designed and applied appropriately in order to prevent any discontinuities of the insulation layer, such as might be caused by utility boxes or external fixings, or introduced through poor workmanship. Any of these, if left unchecked, could compromise the potential effectiveness of applying external insulation to the walls.

4 Detailing of internal wall insulation

4.1 Interstitial condensation risk

When a solid wall is insulated externally, it essentially keeps the wall construction warm. However, the use of internal insulation on solid walls results in much of the wall construction being colder than before. This can be readily seen in the various images of the modelled details that follow, where these show the 'colder' colours applying to the bulk of the wall with only the structure to the inside of the insulation showing the warmest colours. The temperature within the wall drops considerably where the insulation meets the inner wall surface.

Unless moisture is prevented from passing through the insulation layer, eg by the use of a vapour control layer behind the plasterboard, interstitial condensation may occur between the interfaces of (or indeed within) the materials on the cold side of the insulation layer. Before installing internal insulation it is important to assess the condition of each external wall to ensure that it is not damp and that it is adequately protected externally (by render or brickwork pointing in a satisfactory condition) so that moisture will not be able to penetrate into the wall from the outside.

4.2 Window jamb

For the window jamb junction in the 'uninsulated' base case, it is assumed that the window frame is situated in the centre of the wall cross-section, with external render and internal plaster each returned to the window frame (Figures 10a and 10b). For the 'typical' insulation case (Figure 10c), it is assumed that the internally applied insulation will run flush to the inner edge of the jamb and the plaster finish will again be returned to the window.

For the 'improved' insulated case (Figure 10d), a fillet of insulation is applied within the window reveal prior to plastering. For this case, 20 mm of insulation with a conductivity of 0.02 W/mK, such as phenolic foam, is assumed, since typically there will be a 30–40 mm space between the existing wall and the glazing. In reality, the thickness of the applied insulation would need to be tailored to the available space within the reveal.

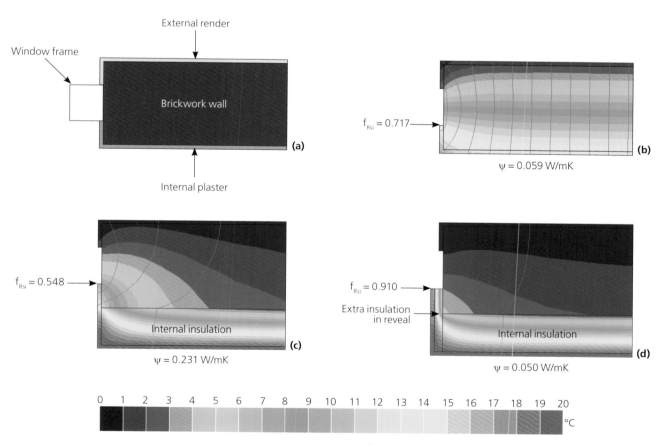

Figure 10: Junction detail for window jamb
a: 'Uninsulated' detail; **b:** 'Uninsulated' modelling results; **c:** 'Typical' internal insulation; **d:** 'Improved' internal insulation

Inward-opening windows present a problem when aiming to improve the detailing of internally applied wall insulation at window reveals and this is often given as the reason why such detailing cannot be practically achieved. If the windows are due for replacement in the near future, it follows that it may be beneficial to consider replacing them at the same time as the internal insulation is applied to the wall, so as to ensure that appropriate window frame arrangements can be selected to allow the additional detail insulation to be included. There may also be scope to insulate across the entire cross-section of the wall at the reveal if the windows are replaced, but this is not explored in this study. This issue regarding replacement windows applies also to the window head and window sill junctions that follow.

For the 'uninsulated' base case (Figures 10a and 10b), the ψ-value is 0.059 W/mK and the minimum temperature factor is 0.717, which is slightly below the critical value of 0.75, indicating that there is some risk of condensation or mould growth. The application of internal insulation (Figure 10c) significantly improves the U-value of the wall and reduces the overall heat loss. The ψ-value of 0.231 W/mK is significantly higher than in the 'uninsulated' base case, the reason being that, although the heat loss through the wall itself has been reduced, there is now more lateral heat flow towards the junction. The concentration of different colour temperature bands around the location of the window frame in Figure 10c shows the rapid temperature drop across this particular location, indicating that lower temperatures are now present at the point where the inner plaster meets the window frame. Consequently, the temperature factor at the coldest internal point is reduced to 0.548, which is well below the critical factor of 0.75, indicating a substantial risk of condensation and mould growth forming at the 'typical' insulation junction.

Figure 10d shows the 'improved' insulation case, where additional insulation is applied to the inside reveal. Here the ψ-value has returned to a more acceptable value of 0.050 W/mK. The internal surface temperature has also risen and the minimum temperature factor has increased to 0.910, now indicating no risk of condensation or mould growth at this 'improved' junction.

4.3   Window head

For the window head junction in the 'uninsulated' base case (Figures 11a and 11b), it is assumed that a concrete lintel is present across the head of the window and that the window frame is again situated in the centre of the wall cross-section, with external render and internal plaster returned to the window frame. For the 'typical' insulation case (Figure 11c), it is assumed that the internal insulation will run flush to the edge of the lintel and the plaster finish will again be returned to the window frame. For the 'improved' insulation case, it is recommended that a fillet of insulation is applied within the window reveal at the head prior to plastering (Figure 11d). In this case, 20 mm of insulation with a conductivity of 0.02 W/mK, such as phenolic foam, is assumed.

For the window head detail and for the 'uninsulated' base case (Figures 11a and 11b), the ψ-value is high at 0.251 W/mK, as a consequence of the concrete lintel being more thermally conductive than the brickwork of the solid wall. The minimum temperature factor of 0.671 is well below the critical value of 0.75 and so there is a significant risk of condensation and mould growth at the internal surface of the window head junction.

The application of 'typical' internal insulation to the solid wall (Figure 11c), ie not insulating the reveal, reduces the overall heat loss, but causes an increase in the ψ-value (compared with the 'uninsulated' base case) to 0.344 W/mK. Again it can be seen that lower temperatures are now present, such that the minimum temperature factor (at the point where the inner plaster meets the window frame) is reduced to 0.509, which is well below the critical factor of 0.75, indicating a considerable risk of condensation and mould growth forming at the newly insulated junction.

Figure 11d shows the 'improved' internal insulation case where insulation is also applied within the inner reveal. Here the ψ-value has reduced to 0.060 W/mK, which is a considerable improvement over both the 'uninsulated' and 'typical' insulation cases. The inside surface temperatures have risen and the minimum temperature factor is much improved to 0.923, indicating no risk of condensation or mould growth at the 'improved' junction. Note that if the concrete lintel were replaced by a traditional timber lintel, the thermal bridging at this junction would be much reduced.

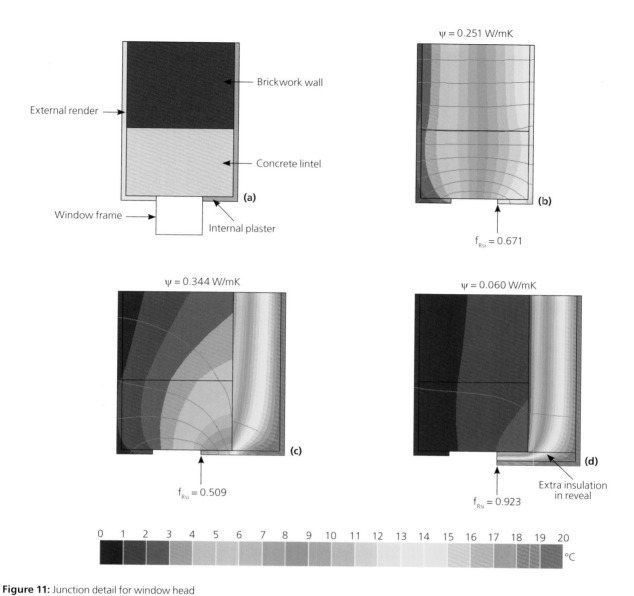

Figure 11: Junction detail for window head
a: 'Uninsulated' detail; **b:** 'Uninsulated' modelling results; **c:** 'Typical' internal insulation; **d:** 'Improved' internal insulation

4.4   Window sill

It is assumed that a traditional limestone sill spans the wall at this junction and that the window frame is again situated in the centre of the wall cross-section. External render is applied up to the underside of the stone sill and a hardwood inner sill board finishes the internal reveal (Figures 12a and 12b). When the wall is internally insulated, it is assumed that the insulation would be brought to the underside of an extended hardwood inner sill board (Figure 12c). To improve this junction, a 20 mm fillet of insulation with a conductivity of 0.02 W/mK, such as phenolic foam, is applied to the stone sill beneath the wooden sill board (Figure 12d).

For the 'uninsulated' base case (Figures 12a and 12b), the ψ-value is 0.186 W/mK and the minimum temperature factor of 0.662 is below the critical value of 0.75, indicating a significant risk of surface condensation and mould growth. For the 'typical' insulation case, the ψ-value increases a little to 0.199 W/mK. However, the minimum temperature factor worsens to 0.585, indicating a high risk of surface condensation and mould growth for this 'typical' insulation case.

For the 'improved' insulation case, additional insulation is applied within the inner reveal above the stone sill and below the inner sill board. As a result, the ψ-value is improved to a more acceptable value of 0.049 W/mK and the minimum temperature factor is 0.891, which is now well above the critical value of 0.75, indicating no risk of surface condensation or mould growth in this case.

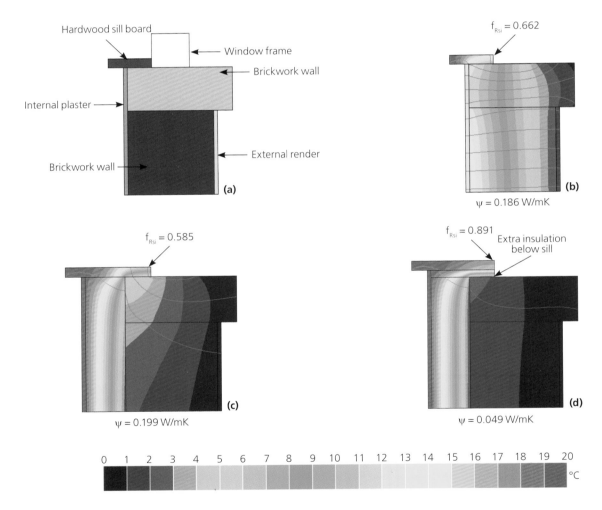

Figure 12: Junction detail for window sill
a: 'Uninsulated' detail; **b:** 'Uninsulated' modelling results; **c:** 'Typical' internal insulation; **d:** 'Improved' internal insulation

4.5 Intermediate floor/ external wall junction (within the same dwelling)

It is assumed that the intermediate floor comprises timber joists tied into the external brick wall, supporting timber floorboards for the upper floor and a lath and plaster-type finish for the ceiling below. This creates a 'narrow', still airspace between the joists, for which appropriate conductivities have been calculated in accordance with BS EN ISO 6946:2007 and following the guidance in BR 497 for the purposes of modelling. The brick wall is externally finished with sand render and plastered internally. A softwood skirting board is also included at the corner of the upper floor junction. For the purposes of modelling according to BR 497, it is not necessary to include the repeating timber joists perpendicular to the external wall when modelling this junction.

For the 'typical' insulation case (Figure 13c), it is assumed that the internal insulation would be applied above and below the existing intermediate floor, with a plaster finish. For the 'improved' insulation case (Figure 13d), it is proposed that the floorboards adjacent to the external wall are removed to allow the insulation to be continued around the joists and so fill the

airspace at the edge of the floor to a thickness corresponding to at least the thermal resistance of the internal insulation thickness applied to the wall.

For the 'uninsulated' base case (Figures 13a and 13b), the ψ-value is, as expected, small and negative at -0.072 W/mK. This is because the area to which the wall U-value is applied includes the area of the edge of the intermediate floor. The minimum temperature factor is 0.708, which is below the critical value of 0.75, indicating a risk of surface condensation and mould growth.

For the 'typical' insulation case (Figure 13c), the application of the internal insulation above and below the intermediate floor reduces the overall heat loss, but the ψ-value is now increased to 0.447 W/mK (as a consequence of the area of the edge of the floor having the 'insulated' U-value applied, when in fact there is no insulation). The minimum temperature factor is 0.821, which is above the critical value of 0.75, indicating no risk of surface condensation or mould growth.

For the 'improved' insulation case, where the edge of the floor is now insulated, the ψ-value is once more negative, but only slightly at -0.002 W/mK. The minimum temperature factor is further improved to 0.958, indicating again no risk of surface condensation or mould growth.

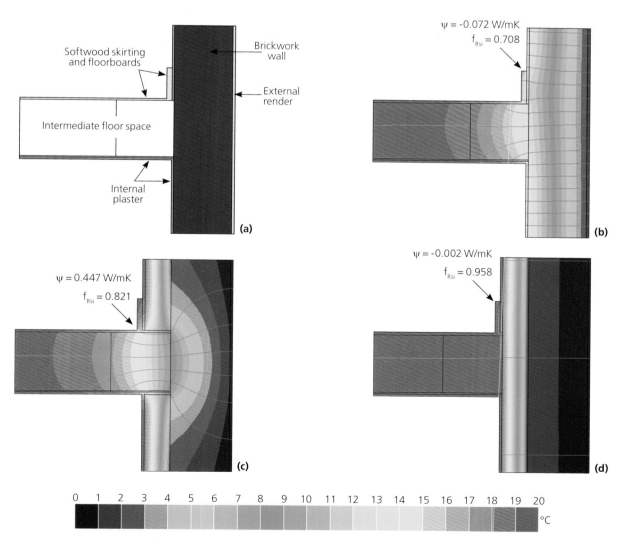

Figure 13: Junction detail for intermediate floor/external wall junction (within the same dwelling)
a: 'Uninsulated' detail; **b:** 'Uninsulated' modelling results; **c:** 'Typical' internal insulation; **d:** 'Improved' internal insulation

4.6 Intermediate floor/ external wall junction (in apartments)

There may be instances, typically within apartments, where internally applied insulation is only applied to one storey (ie within one apartment) rather than across the entire wall. The effect of this occurrence is explored here to demonstrate the likely impact on heat flows and surface temperature factors for each apartment and how the resultant ψ-value would be attributed to the junction in such instances. However, it is noted that there is no feasible detailing solution likely to mitigate this effect, short of ensuring that all apartments are insulated internally, with attention paid to insulating within the intermediate floor, as discussed in Section 4.5.

As before, it is assumed that the intermediate floor would comprise timber joists tied into the external brick wall, supporting timber floorboards for the upper floor and a double thickness of staggered-fit, fire-resistant plasterboard finish to the ceiling below. This creates a 'narrow', still airspace between the joists, for which appropriate conductivities have been calculated in accordance with BS EN ISO 6946:2007 and following the guidance in BR 497 for the purposes of modelling. The brick wall is externally finished with sand render and plastered internally. A softwood skirting board is also included for the upper floor of the junction. Again, for the purposes of modelling according to BR 497, it is not necessary to include the repeating timber joists perpendicular to the external wall when modelling this junction. For the insulated case (Figure 14c), it is assumed that internal insulation would be installed in the upper apartment, running flush to the existing floor line with a plaster finish applied.

When calculating the heat loss for these apartments, the U-values are measured to the internal finished surfaces of the ceiling and floor of each apartment respectively. Hence, in order for the ψ-value to represent the residual heat loss from the junction, it includes the heat loss across the area of wall adjacent to the intermediate floor space. Even in the uninsulated base case scenario, the heat flow across the junction is asymmetric, at 0.156 W/mK for the upper apartment and 0.274 W/mK for the lower apartment. In both apartments, the temperature factor is below the critical value of 0.75, indicating a risk of surface condensation and mould growth at the inner surfaces of the junction in both apartments.

Where the upper apartment is insulated and the lower apartment is not (Figure 14c), heat loss from the insulated dwelling will be much reduced compared with the base case. As a consequence, however, the heat flow across the junction from the upper apartment compared with the lower apartment is now very asymmetric, with the insulated apartment having a relatively unchanged ψ-value of 0.159 W/mK and that of the uninsulated apartment an increased ψ-value of 0.351 W/mK.

The temperature factor for the insulated apartment rises above the critical factor when internal insulation is applied, hence there would no longer be a risk of condensation or mould growth at the floor junction of this apartment. However, the temperature factor on the uninsulated side is further reduced below the critical factor down to 0.707, indicating that the risk of surface condensation or mould growth has increased slightly in the uninsulated dwelling. As already noted, the only way to mitigate this effect would be to ensure that all apartments are insulated internally, with attention paid to insulating within the intermediate floor, as discussed in Section 4.5.

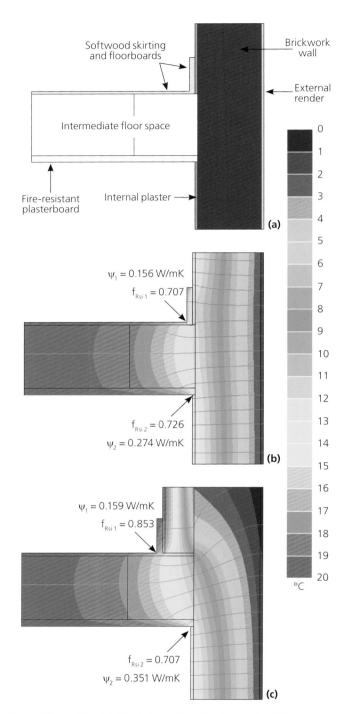

Figure 14: Junction detail for intermediate floor/external wall junction (in apartments)
a: 'Uninsulated' detail
b: 'Uninsulated' modelling results
c: Internal insulation (upper apartment)

4.7   Party wall/external wall junction

Since the majority of dwellings in the UK are attached to a neighbouring dwelling (ie not detached), it is relatively likely that situations will arise when one property in a neighbouring pair will receive internally applied wall insulation, but the other will not.

It is assumed for the 'uninsulated' base case that the external wall and the adjoining party wall are of brickwork construction, with an internal plaster finish and an external sand render finish (Figures 15a and 15b). For the 'typical' insulation case, where only one of the adjoining dwellings receives internally applied insulation, it is assumed that the internal insulation will be taken flush to the party wall and finished with plaster (Figure 15c). For the 'improved' insulation case, the frequently suggested improvement of returning the insulation along the party wall is modelled (Figure 15d).

For the 'uninsulated' base case, the ψ-value is 0.465 W/mK. Since the heat flows across the junction are symmetrical, half of this is attributed to each property. However, in the 'typical' insulation case where one property is insulated and the other not (Figure 15c), heat loss from the insulated dwelling will be much reduced compared with the base case, while that for the

uninsulated dwelling is increased. As a consequence, the heat flow across the junction from one property compared with the other is now very asymmetric. Because of this, separate ψ-values for this junction must be determined for each property. In this case, instead of the total ψ-value (0.488 W/mK) being halved, the insulated property has a ψ-value of 0.169 W/mK and that of the uninsulated property a value of 0.319 W/mK.

Note that the temperature factor of the base case (0.730) is just below the critical value (0.75), but it is lowered further after the 'typical' internal insulation has been installed in the adjacent dwelling. The temperature factor on the uninsulated side is reduced below the critical factor to 0.697 and the ψ-value of the junction as a whole is increased slightly. The risk of condensation therefore has increased for the uninsulated dwelling.

When insulation is returned along the party wall on the insulated side (Figure 15d), the temperature factor drops even further for the uninsulated neighbouring dwelling (to 0.686). Evidently, part of the bridging effect at this junction is that the heat flows towards the uninsulated property as a means of bypassing the internally applied insulation, thus marginally benefiting the uninsulated dwelling compared with when the heat loss is further isolated by returning insulation along the party wall. Hence, this does not appear a particularly worthwhile measure on the whole.

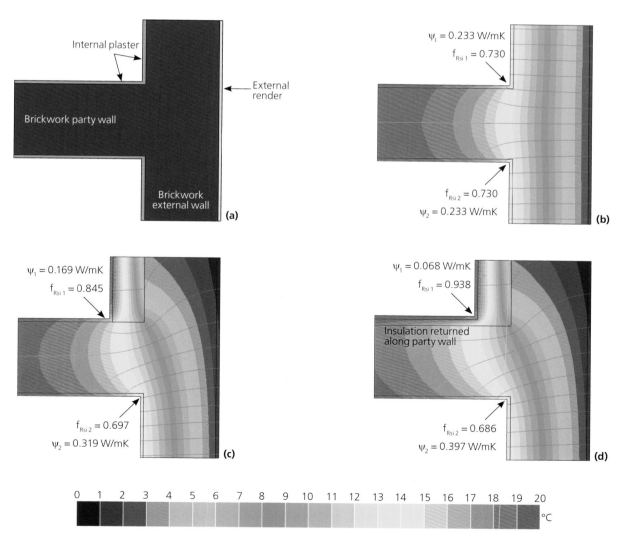

Figure 15: Junction detail for party wall/external wall junction (only one property insulated)
a: 'Uninsulated' detail; **b:** 'Uninsulated' modelling results; **c:** 'Typical' internal insulation; **d:** Internal insulation returned along party wall

It is worth noting that the thermal bridging at this junction would not be significantly improved even if both properties were insulated. If this is done, the ψ-value for the whole junction is reduced from the baseline total of 0.466 W/mK to 0.458 W/mK, with 0.229 W/mK allocated to each dwelling.

4.8   Quality control on site

When fitting internal wall insulation, many practical detailing considerations need to be taken into account. It is necessary to ensure that further unintended thermal bridges are not built into the structure by poor workmanship or inadequate detailing of junctions/connections other than those discussed in this study. This section highlights examples of such instances, as witnessed in real site installations.

4.8.1 Workmanship

Additional thermal bridges can be introduced by poor workmanship that causes inconsistency in the applied insulation. The most significant example is poor butting-up/alignment of insulation creating small gaps in the insulation layer. Care should be taken when detailing around openings, particularly where the type of insulation changes at reveals, to ensure no unintended gaps are introduced that could negate the benefit of the improved detail being delivered. Although such gaps may be sealed to prevent potential moisture ingress, such sealants will generally not contribute to the insulation of the wall. Hence, areas where heat can bypass the insulation may occur. An individual instance of this is unlikely to result in significant heat loss from the building, but poor installation in a number of places across a dwelling could do so collectively.

4.8.2 Ventilation considerations

While sealing gaps to reduce draughts may not offer resistance to the transfer of heat, it is very important to seal the newly applied insulation layer to prevent the passage of moisture behind the insulation at junctions, which could potentially lead to interstitial condensation on the colder inner surface of the original wall (as discussed in Section 4.1). At the same time, consideration must also be given to the controlled ventilation of the newly insulated building (including adequate ventilation in wet rooms through extract fans), since the insulation will effectively create a new barrier to the passage of moisture.

4.8.3 Obstructions and detailing

As with obstructions hindering the installation of external wall insulation, such as utility boxes or adjoining building features, it is important to minimise disruptions to the continuity of internally applied insulation. The number of such obstacles on internal walls can be considerable, including radiators, pipe runs, plug and phone sockets, light switches, skirting boards and architraves. It would be tiresome and problematic to try to cut insulation to fit around all of these items, so from a practical and thermal bridging perspective it is better to remove them from the wall if possible and subsequently refit them to allow the insulation to continue uninterrupted.

In addition, it should be noted that any instances where the insulation is not continuous will allow heat to bypass the insulation layer and become concentrated areas of heat loss. More significantly, this may result in a relative temperature drop in these areas and the temperature factor may fall below the critical limiting value. This could lead to an increased risk of condensation and mould growth in the uninsulated areas.

Therefore, prior to the commencement of any internal insulation works, potential obstructions and irregular detailing should be identified and measures to eliminate such problem areas should be proposed.

Additional and significant obstacles include kitchen and bathroom units, which are often fixed to external walls. Due to the extent of the relocation works that would be required in such instances, it is generally recommended that internal insulation measures are phased to coincide with the replacement of kitchen or bathroom units. However, where such phasing is not feasible, it should not be considered acceptable to leave hidden segments of the wall uninsulated. Another point to consider is the selection of the fixings needed to enable the hanging of kitchen or bathroom units from an internally insulated wall, which is likely to necessitate reinforced battening in key locations (with consideration for future layout changes). It should be recognised that large gauge metal fixings in such situations would introduce their own 'point' thermal bridges (calculated as χ-values) contributing to additional heat loss from the building, since metals are relatively conductive.

4.9   Conclusions for internal wall insulation

Modelling these examples of key junctions for internally applied insulation demonstrates that it is possible for thermal bridging to increase after internal insulation has been applied to solid walls, when typical industry practices are followed. However, these increased thermal bridging effects can generally be mitigated by giving special attention to the methods of detailing employed. In designing the detailing around junctions when installing internal insulation, the following should be considered:

- Insulate within the reveals at window junctions; in particular, isolate thermally conducting features such as concrete or stone lintels or sills.

- Where external walls meet a timber intermediate floor, it is possible to reduce the extent of thermal bridging by removing the floorboards adjacent to the external wall and the corresponding width of ceiling below to allow the insulation to be continued around the joists up the entire inner surface of the wall.

At the junction of party walls with external walls, the extent of thermal bridging cannot be significantly reduced, even when both neighbouring dwellings are insulated internally. The modelling has further shown that returning insulation up to 1 m along the party wall of the internally insulated dwellings provides only a limited benefit, and that this has the disadvantage of marginally increasing the condensation risk in an uninsulated neighbouring dwelling. Hence returning the insulation along the party wall does not appear to be a worthwhile step on the whole.

For the majority of the models, the temperature factor is made worse as a result of poorly detailed internally applied insulation, thus increasing the risk of condensation compared with the base case. However, this risk can be reduced when additional detailing measures are taken, as described in this report.

It is important to ensure that internal wall insulation is designed and applied appropriately to ensure that no unintended gaps are introduced through poor design or workmanship to compromise the potential effectiveness of the internally applied insulation.

5 Comparing the overall thermal performance of external and internal insulation

To help understand and compare the overall thermal benefit of insulating solid walls with either external or internal insulation, the relevant ψ-values for the key external and internal junctions have been applied, along with the relevant U-values for the plane elements of the building fabric, to a 'theoretical' semi-detached (or end-terrace) two-storey dwelling whose dimensions are 5 m wide, 8 m deep and 5 m high to the eaves. Overall, the area of openings is approximated at 11.4 m². In addition to the junction details discussed in this report, further modelling has been carried out on the remaining junctions that would be present in this example dwelling, so as to provide a proper comparison. The various heat losses are calculated for the plane elements (U-values × areas) and junctions (ψ-values × lengths) and summed. This then enables comparisons to be made between insulating externally or internally. The detailed heat losses are given in Table 3. When this is compared with Tables 4 and 5 for the dwellings with typically applied external and internal insulation, the energy loss from the junctions is at least doubled for internal insulation and higher again for externally applied insulation. However, the overall energy loss from the fabric and junctions combined is significantly reduced due to the improvement in the U-value of the insulated walls. As demonstrated in the earlier sections of this report, when the applied insulation is not detailed effectively at key junctions, heat losses can actually increase for these junctions. This leads to the thermal bridging being a greater percentage of the total heat losses relative to the uninsulated case. The improved detailing demonstrated in this report can cut the increased heat loss from certain junctions and in the case of internally applied insulation can actually reduce the total junction heat losses to below those of the uninsulated base case, as shown in Tables 6 and 7. This, in combination with the reduced heat loss from the fabric elements due to the wall insulation, leads to a large reduction in heat loss from the example building overall.

5.1 Conclusions for external versus internal wall insulation of solid walls

Often the decision whether to install internal or external insulation to solid walls is driven by practical issues and key features of the dwelling, as discussed in Section 2. It is interesting to note, however, that the total heat loss from junctions with internally applied insulation is slightly lower than the heat loss from junctions with externally applied insulation, though this gap is narrowed where the improved junction detailing scenarios are used to provide further reductions in heat loss. Realistically, the difference is not substantial enough between external and internally applied insulation to particularly favour one over the other on the grounds of thermal bridging alone. Hence, it seems likely that the decision between choosing internal or external insulation will continue to be based on practical factors relating to the particular building as a whole.

Table 3: Heat loss in example dwelling (uninsulated)

Junction	Length (m)	ψ-value (W/mK)	Total (W/K)
Ground floor/external wall	20.0	0.149	2.98
Internal floor/external wall	20.0	-0.072	-1.44
Eaves	12.0	-0.012	-0.14
Gable	8.0	0.160	1.28
Head	8.0	0.251	2.01
Sill	8.0	0.186	1.49
Jamb	22.8	0.059	1.35
Corner	10.0	0.192	1.92
Party wall/external wall	10.0	0.233	2.33
Ground floor/party wall	8.0	-0.057	-0.46
Party wall/roof	8.0	0.179	1.43
Total heat loss from junctions			**12.73**

Fabric element	Area (m²)	U-value (W/m²K)	Total (W/K)
Walls	88.6	1.985	175.87
Roof	48.0	0.201	9.65
Floor	48.0	0.398	19.10
Windows	11.4	2.200	25.08
Total heat loss from fabric elements			**229.70**
Overall heat loss			242.44
Junction loss as % of fabric loss			5.54

Table 4: Heat loss in example dwelling with typically applied external insulation

Junction	Length (m)	ψ-value (W/mK)	Total (W/K)
Ground floor/external wall	20.0	0.604	12.09
Internal floor/external wall	20.0	-0.001	-0.02
Eaves	12.0	0.331	3.97
Gable	8.0	0.258	2.06
Head	8.0	0.445	3.56
Sill	8.0	0.483	3.86
Jamb	22.8	0.310	7.08
Corner	10.0	0.112	1.12
Party wall/external wall	10.0	0.329	3.29
Ground floor/party wall	8.0	-0.057	-0.46
Party wall/roof	8.0	0.179	1.43
Total heat loss from junctions			**37.99**

Fabric element	Area (m²)	U-value (W/m²K)	Total (W/K)
Walls	88.6	0.261	23.12
Roof	48.0	0.201	9.65
Floor	48.0	0.398	19.10
Windows	11.4	2.200	25.08
Total heat loss from fabric elements			**76.96**
Overall heat loss			114.94
Junction loss as % of fabric loss			49.36

Table 5: Heat loss in example dwelling with typically applied internal insulation

Junction	Length (m)	ψ-value (W/mK)	Total (W/K)
Ground floor/external wall	20.0	0.293	5.86
Internal floor/external wall	20.0	0.447	8.94
Eaves	12.0	0.028	0.34
Gable	8.0	0.041	0.33
Head	8.0	0.344	2.75
Sill	8.0	0.199	1.59
Jamb	22.8	0.231	5.26
Corner	10.0	0.021	0.21
Party wall/external wall	10.0	0.169	1.69
Ground floor/party wall	8.0	-0.057	-0.46
Party wall/roof	8.0	0.179	1.43
Total heat loss from junctions			**27.94**

Fabric element	Area (m²)	U-value (W/m²K)	Total (W/K)
Walls	88.6	0.261	23.12
Roof	48.0	0.201	9.65
Floor	48.0	0.398	19.10
Windows	11.4	2.200	25.08
Total heat loss from fabric elements			**76.96**
Overall heat loss			104.89
Junction loss as % of fabric loss			36.30

Table 6: Heat loss in example dwelling with external insulation and improved detailing at key junctions

Junction	Length (m)	ψ-value (W/mK)	Total (W/K)
Ground floor/external wall	20.0	0.154	3.08
Internal floor/external wall	20.0	-0.001	-0.02
Eaves	12.0	0.042	0.50
Gable	8.0	0.258	2.06
Head	8.0	0.064	0.51
Sill	8.0	0.110	0.88
Jamb	22.8	0.056	1.28
Corner	10.0	0.112	1.12
Party wall/external wall	10.0	0.329	3.29
Ground floor/party wall	8.0	-0.057	-0.46
Party wall/roof	8.0	0.179	1.43
Total heat loss from junctions			**13.68**

Fabric element	Area (m²)	U-value (W/m²K)	Total (W/K)
Walls	88.6	0.261	23.12
Roof	48.0	0.201	9.65
Floor	48.0	0.398	19.10
Windows	11.4	2.200	25.08
Total heat loss from fabric elements			**76.96**
Overall heat loss			90.64
Junction loss as % of fabric loss			17.78

Table 7: Heat loss in example dwelling with internal insulation and improved detailing at key junctions

Junction	Length (m)	ψ-value (W/mK)	Total (W/K)
Ground floor/external wall	20.0	0.293	5.86
Internal floor/external wall	20.0	-0.002	-0.05
Eaves	12.0	0.028	0.34
Gable	8.0	0.041	0.33
Head	8.0	0.060	0.48
Sill	8.0	0.049	0.39
Jamb	22.8	0.050	1.15
Corner	10.0	0.021	0.21
Party wall/external wall	10.0	0.169	1.69
Ground floor/party wall	8.0	-0.057	-0.46
Party wall/roof	8.0	0.179	1.43
Total heat loss from junctions			**11.36**

Fabric element	Area (m²)	U-value (W/m²K)	Total (W/K)
Walls	88.6	0.261	23.12
Roof	48.0	0.201	9.65
Floor	48.0	0.398	19.10
Windows	11.4	2.200	25.08
Total heat loss from fabric elements			**76.96**
Overall heat loss			88.32
Junction loss as % of fabric loss			14.77

6 References

1. Ward T I. Assessing the effects of thermal bridging at junctions and around openings. BRE IP 1/06. Bracknell, IHS BRE Press, 2006.

2. Pearson C J. The complete guide to external wall insulation. York, Wellgarth Publishing Ltd, 2009, 2nd edn.

3. Ward T and Sanders C. Conventions for calculating linear thermal transmittance and temperature factors. BRE BR 497. Bracknell, IHS BRE Press, 2007.

4. BSI. Building components and building elements – Thermal resistance and thermal transmittance – Calculation method. BS EN ISO 6946:2007. London, BSI, 2007.

Other reports from BRE Trust

Modern methods of house construction: a surveyor's guide. **FB 11**

Crime opportunity profiling of streets (COPS): a quick crime analysis – rapid implementation approach. **FB 12**

Subsidence damage to domestic buildings: a guide to good technical practice. **FB 13**

Sustainable refurbishment of Victorian housing: guidance, assessment method and case studies. **FB 14**

Putting a price on sustainable schools. **FB 15**

Knock it down or do it up? **FB 16**

Micro-wind turbines in urban environments: an assessment. **FB 17**

Siting micro-wind turbines on house roofs. **FB 18**

Automatic fire sprinkler systems: a guide to good practice. **FB 19**

Complying with the Code for Sustainable Homes: lessons learnt on the BRE Innovation Park. **FB 20**

The move to low-carbon design: are designers taking the needs of building users into account? **FB 21**

Building-mounted micro-wind turbines on high-rise and commercial buildings. **FB 22**

The real cost of poor housing. **FB 23**

A guide to the Simplified Building Energy Model (SBEM): what it does and how it works. **FB 24**

Vacant dwellings in England: the challenges and costs of bringing them back into use. **FB 25**

Energy efficiency in new and existing buildings: comparative costs and CO_2 savings. **FB 26**

Health and productivity benefits of sustainable schools: a review. **FB 27**

Integrating BREEAM throughout the design process: a guide to achieving higher BREEAM and Code for Sustainable Homes ratings through incorporation with the RIBA Outline Plan of Work and other procurement routes. **FB 28**

Design fires for use in fire safety engineering. **FB 29**

Ventilation for healthy buildings: reducing the impact of urban pollution. **FB 30**

Financing UK carbon reduction projects. **FB 31**

The cost of poor housing in Wales. **FB 32**

Dynamic comfort criteria for structures: a review of UK standards, codes and advisory documents. **FB 33**

Water mist fire protection in offices: experimental testing and development of a test protocol. **FB 34**

Airtightness in commercial and public buildings. 3rd edn. **FB 35**

Biomass energy. **FB 36**

Environmental impact of insulation. **FB 37**

Environmental impact of vertical cladding. **FB 38**

Environmental impact of floor finishes: incorporating The Green Guide ratings for floor finishes. **FB 39**

LED lighting. **FB 40**

Radon in the workplace. 2nd edn. **FB 41**

U-value conventions in practice. **FB 42**

Lessons learned from community-based microgeneration projects: the impact of renewable energy capital grant schemes. **FB 43**

Energy management in the built environment: a review of best practice. **FB 44**

The cost of poor housing in Northern Ireland. **FB 45**

Ninety years of housing, 1921–2011. **FB 46**

BREEAM and the Code for Sustainable Homes on the London 2012 Olympic Park. **FB 47**

Saving money, resources and carbon through SMARTWaste. **FB 48**

Concrete usage in the London 2012 Olympic Park and the Olympic and Paralympic Village and its embodied carbon content. **FB 49**

A guide to the use of urban timber. **FB 50**

Low flow water fittings: will people accept them? **FB 51**

Evacuating vulnerable and dependent people from buildings in an emergency. **FB 52**

Refurbishing stairs in dwellings to reduce the risk of falls and injuries. **FB 53**

Dealing with difficult demolition wastes: a guide. **FB 54**

Security glazing: is it all that it's cracked up to be? **FB 55**

The essential guide to retail lighting. **FB 56**

Environmental impact of metals. **FB 57**

Environmental impact of brick, stone and concrete. **FB 58**

Design of low-temperature domestic heating systems. **FB 59**

Performance of photovoltaic systems on non-domestic buildings. **FB 60**

BRE Connect Online
Build on your foundation of **knowledge** and **expertise**

WHAT IS BRE CONNECT ONLINE?

BRE Connect Online gives you **access to the unrivalled expertise and insight of BRE** – the UK's leading centre of excellence on the built environment. BRE Connect Online is an **annual subscription service** from IHS BRE Press giving online access to **over 1700 BRE titles**

WHAT DO I GET?

ALL NEW AND PUBLISHED BRE TITLES

700 books, reports and guides – research, innovation, best practice and case studies, including:

- The Green Guide to Specification
- Designing quality buildings
- Foundations, basements and external works
- Roofs and roofing
- Site layout planning for daylight and sunlight

250 Digests – authoritative state-of-the-art reviews

600 Information Papers – BRE research and how to apply it in practice

150 Good Building and Repair Guides – illustrated practical guides to good building and repair work

AND MUCH MORE ...

WHAT'S NEW IN 2013?

MORE THAN 50 NEW TITLES, INCLUDING:

- Delivering energy efficiency in commercial buildings
- Design of low-temperature domestic heating systems
- Energy surveys and audits
- External fire spread performance of cladding systems on multi-storey buildings
- Performance of photovoltaic systems on non-domestic buildings
- Radon solutions in homes
- Security glazing
- The essential guide to retail lighting

All this for an annual subscription of only **£399** + VAT
Call now on +44 (0) 1344 328038 to find out more or arrange a free trial